Megalodon Is True
Ancient king of the oceans

AZEDDINE BELLOUK

ISBN: 9798377462347

Table of Contents

Introduction

Welcome to Creature Scene Investigation: The Science of Cryptozoology, the series dedicated to the science of cryptozoology. physiologist Heuvelmans, a French Scientists fictional that word fifty years ago. it's a mixture of the words kryptos (Greek for "hidden") and zoology, the scientific study of animals. So, cryptozoology is that the study of "hidden" animals, or cryptids, that are animals that some individuals believe might exist, even if it is not nevertheless proven. simply however will someone prove that a selected cryptid exists? Dedicated cryptozoologists (the scientists who study cryptozoology) follow a long, ballroom dance method as they search for cryptids. First, they gather the maximum amount data regarding their animals as they can. the foremost necessary sources of information are folks that live close to wherever the cryptid purportedly lives. These individuals are most accustomed to the animal and therefore the stories about it. So, for example, if cryptozoologists want to search out out about the Loch Ness Monster, they need to raise the people who live around Loch Ness, a lake in Scotland where the monster was sighted.

If they need to be told about Bigfoot, they should consult with folks that found its footprints or took its photo. A cryptozoologist rigorously examines all of this data. this {can be} necessary as a result of it helps the person determine and rule out some stories that may be mistakes or lies.

The remaining information can then be wont to turn out a transparent scientific description of the cryptid in question. it would even result in solid proof that the cryptid exists. Second, a cryptozoologist takes the results of his or her analysis and goes into the sphere to appear for solid proof that the cryptid extremely exists. the simplest potential proof would be associate degree actual specimen—maybe even a live one. in need of that, a combination of excellent videos, photographs, footprints, body components (bones and teeth, for example), and different clues will create a powerful case for a cryptid's existence.

during this way, the science of cryptozoology may be a heap like forensics, the science created famed by all of these crime investigation shows on TV. The goal of forensics detectives is to use the evidence they notice to catch a criminal. The goal of cryptozoologists is to catch a cryptid—or a minimum of to search out solid proof that it extremely exists.

Some cryptids became world-famous. the foremost famed ones of all are most likely the legendary Loch Ness Monster of Scotland and therefore the apelike Sasquatch of the United States. There are several different cryptids out there, too. At least, some individuals suppose so.

This collection explores the legends and lore—the records and the fiction in the back of the maximum famous of all the cryptids:

the large shark called Megalodon, Kraken the monster squid, an African dinosaur referred to as Mokele-mbembe, the Loch Ness Monster, and Bigfoot. This collection additionally takes a look at a few lesser-regarded however similarly captivating cryptids from across the world:

• the mysterious, blood-sucking Chupacabras, or "goat sucker," from the Caribbean, Mexico, and South America

• the Sucuriju, a massive anaconda snake from South America

• Megalania, the large reveal lizard from Australia

• the Ropen and Kongamato, prehistoric flying reptiles from Africa and the island of New Guinea

• the thylacine, or Tasmanian wolf, from the island of Tasmania

- the Ri, a mermaidlike creature from the waters of New Guinea
- the thunderbird, a massive vulture from western North America

Some cryptids, inclusive of dinosaurs like Mokele-mbembe,
are animals already regarded to technology. These animals are notion to have turn out to be extinct. Some people, however, consider that those animals are nevertheless alive in lands that are tough for maximum people to reach. Other cryptids, inclusive of the massive anaconda snake, are really surprisingly large (or, in a few cases, surprisingly small) variations of current animals.

And but different cryptids, inclusive of the Chupacabras, seem to be animals proper out of a technology fiction movie, completely unlike whatever regarded to fashionable technology.

As cryptozoologists look for those uncommon animals, they preserve in thoughts more than one slogans. The first is, "If it sounds too precise to be true, it in all likelihood isn't true." The second is, "Absence of evidence isn't always evidence of absence." The which means of those slogans turns into clean as you take a look at how cryptozoologists examine and interpret the proof they accumulate in their look for those notable animals

The nightmare of the fisherman: Megalodon

The truly amazing, mind-blowing thing is to imagine —and it might be true—that there are great whites that are 100 feet long deep within the ocean. Carcharodon carcharias is the Latin name for this fish, okay? The closest living relative we can identify is a fish called Carcharodon megalodon, which lived about 30,000–40,000 years ago. We have Megalodon fossilized teeth. They have a six-inch length.

The fish would be between 80 and 100 feet deep in that case. And the teeth are identical to those in modern great whites. To illustrate, let's say the two fish are actually members of the same species.

What makes us believe Megalodon is indeed extinct? Why is it the case? not a food shortage. If there is sufficient water down there to support whales, then there is sufficient water to support large sharks. The fact that we haven't seen a 100-foot white doesn't rule out the possibility that they could exist. What could it accomplish and what kind of power would it possess?

It would resemble a locomotive with butcher knives jammed into its mouth.

Great white shark, or "white shark," Strange, huh? Any beachgoer would hesitate before wading into the sea after hearing these three ominous words. That's mostly because of another short word: Jaws. This well-known book by Peter Benchley, which was adapted into a hugely successful film in 1975, tells the horrifying story of a great white shark terrorizing the locals of Amity, New York, a sleepy little vacation destination on the Atlantic coast. Anyone can be persuaded by this thrilling tale of man versus a man-eating shark that great white sharks are among the most dreadful predators that have ever swum the seven seas.

One potent predator is the great white shark. This enormous fish swims at the top of the ocean food chain and can reach lengths of more than 20 feet (about 6 meters) and weights of more than two tons (1,800 kilograms). This shark hunts fish, pinnipeds (seals and sea lions), dolphins, and porpoises while cruising cool coastal waters between the warm tropics and the icy polar seas.

One of the ocean's most dreadful predators is the great white shark. Great whites can grow to be more than 20 feet (6 meters) long and about 5,000 pounds in weight (2,260 kg). Human attacks by great white sharks have been documented, however they are extremely uncommon.

When a great white shark attacks, it dispatches its prey quickly. A great white once attacked a 200-pound (91 kg), 6-foot (1.8 m) long harbor seal, according to a marine biologist (a scientist who studies life in the ocean). It only took the shark five minutes to kill and devour its prey. The seal vanished after taking three sizable bites.

Great whites as long as 20 feet (6 meters) have been caught by deep sea anglers using a rod and reel. Imagine the trophy that such a monster's jaws would make: a mouth that is 1.5 meters (0.5 feet) wide, lined with 2 inches (5 cm) long, knife-sharp triangular teeth, and large enough to fit your entire head inside. You wouldn't really want to because those teeth are incredibly sharp. A fish that large, powerful, and hazardous could only be safely caught by a very trained angler.

Consider what it would be like to hook a great white that was so large that its teeth measured 6 inches (15 cm), dwarfing the relatively insignificant 2-inchers found on a 20-foot shark. That would be one enormous monster. A whale-sized white shark with a mouth large enough to eat basketball player Shaquille O'Neal in one bite. A fisherman's worst nightmare would undoubtedly be to capture such a beast. Is it possible that such a monster exists? Yes.

Introduction Megalodon

The name given by scientists to this enormous great white shark is Carcharodon megalodon. Numerous Carcharodon megalodon fossilized teeth have been found and painstakingly extracted from sedimentary rocks throughout the years. Megalodon, which is Greek for "mighty tooth," is the common name for this dinosaur. Many more have been rescued from the ocean's depths, numbering in the hundreds. Largest Megalodon tooth ever discovered was enormous, measuring 6.8 inches (17.3 cm) in length. The size of a man's hand.

Even though Megalodon teeth are significantly larger than white shark teeth, they are nonetheless saw-toothed triangles with serrations on both sides. In fact, ichthyologists (scientists who study fish) are confident that the sharks themselves must be identical because the teeth of these two species of sharks are so similar. Megalodon has even been compared as a giant great white shark by some individuals. We won't know for certain until one of these leviathans is captured.

It should be noted that a lot of scientists think the white shark and Megalodon are two species that belong to the same genus, Carcharodon. Although other shark specialists disagree, they believe that because the teeth of these two fishes are sufficiently dissimilar (Megalodon teeth have a scar-like feature near the base and have smaller serrations around the margins), Megalodon should be classified as a new genus, Carcharocles. The recent discovery of a rare, entire fossil jaw belonging to a great white shark ancestor supports this view. This 5-million-year-old fossil's tooth size and arrangement suggest that the great white and another man-eater, the mako shark (genus Isurus), may be more closely related than Megalodon. Whatever the case, most experts concur that Megalodon and the great white are highly similar and likely had a similar appearance.

It is thought that this Megalodon tooth is 5 million years old. Megalodon teeth have been discovered all over the world, indicating that the shark was once quite common in the oceans of the world.

Unfortunately, the majority of shark specialists are pessimistic about ever catching a live Megalodon. Given that no one has ever handled a large shark, either alive or dead, they believe it to be extinct. Nevertheless, a very accurate depiction of Megalodon's appearance and lifestyle has been created by experts. They've been able to do this by researching the remains of extinct sharks and by examining the characteristics and habits of contemporary sharks, particularly the great white, the smaller cousin of the Megalodon. The picture that scientists have drawn is, to put it mildly, breathtaking.

Rebuilding Mighty Tooth

Most people wouldn't think that researching Megalodon's teeth could provide much information for someone trying to reconstruct it, but that is exactly what researchers have done. Scientists have a reasonably solid notion of what the jaws of a shark sporting such lethal dentures by comparing fossilized megalodon teeth with the teeth and jaws of great whites and other current and extinct sharks. They are aware of the teeth's size, form, and other physical characteristics, including how thick and robust they would be.

The considerably smaller great white shark is depicted above a megalodon shark (Carcharodon megalodon) (Carcharodon carcharias). It is estimated that the enormous Megalodon existed between 20 million and 1.2 million years ago.

Scientists believe that because Megalodon teeth are larger and more robust than those of a white shark, Megalodon's jaws would have to be larger and more robust than those of a white shark in order to hold and sustain all those gigantic teeth. This would suggest that the remainder of the skull—and subsequently the entire head—would likewise be larger. Because of this data, Megalodon's pectoral fins should have been bigger and more durable to support and guide the shark's "top-heavy" front end through the water. R. Aldan Martin, a shark researcher, describes the total impact as "kind of a great white on steroids" on his Biology of Sharks and Rays Web site.

How much did Megalodon expand in size? A replica of a Megalodon's jaws was shown at the American Museum of Natural History in New York City in the early 1900s.
The fossil Megalodon teeth utilized in this model came from a variety of sources (and, therefore, from several different sharks). The museum's artists created a set of jaws that were 10 feet (3 m) wide using the number of teeth found in a great white shark's jaws (approximately 24 in each jaw, omitting the hundreds of reserve teeth waiting to replace those that break off). With this enormous jaw, a Megalodon would have been nearly to 30 meters (100 feet) long!

This model comes close to representing the megalodon jaw's size. Undoubtedly, Megalodon was among the sea's most formidable predators.

LET's GET TECHNICAL: EXTERNAL ANATOMY OF SHARKS

During the course of this investigation, it will be necessary to compare Megalodon with other shark species. In order to do this effectively, it is necessary to be familiar with some of the visible body parts of sharks. With that in mind, let's take a whirlwind tour of the external anatomy of the shark that probably looks the most like Megalodon: *Carcharodon carcharias*, the great white shark.

At the business end of the white shark is the tapered, pointed snout, the equivalent of a nose. Like most noses, the white shark's nose has a pair of nostrils, located on the **ventral** (bottom) surface

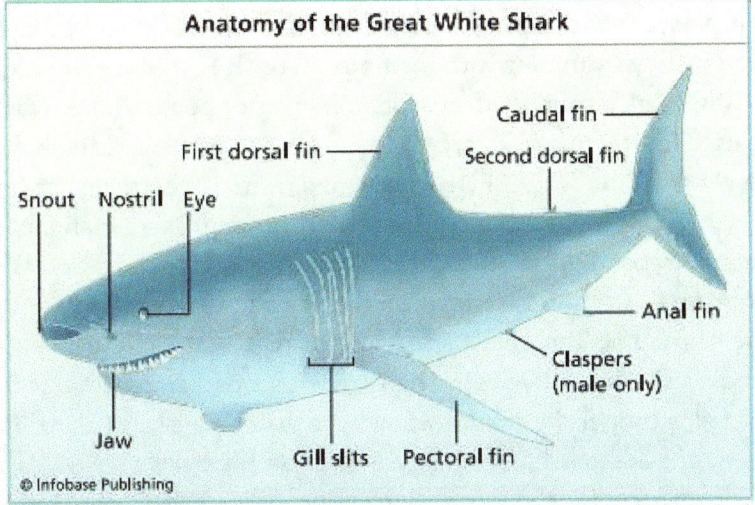

Anatomy of the Great White Shark

Caudal fin

First dorsal fin — Second dorsal fin

Snout   Nostril   Eye

Anal fin

Claspers (male only)

Jaw

Gill slits   Pectoral fin

© Infobase Publishing

The anatomy of the great white shark is perfectly designed for strength and speed. The anatomy of Megalodon was likely very much like that of the great white.

of the snout, a short distance in back of the snout tip. The eyes are located a little further back, on the sides of the head. The mouth, with its tooth-studded jaws, is located directly beneath the eyes, on the ventral surface of the head. Along each side of the shark's neck are five parallel gill slits, where water that has passed through the mouth and gills (which take up oxygen) exits the body. (Many sharks also have spiracles—small openings behind the eyes—that allow water to flow through the gills. Fast swimmers such as the great white often do not have spiracles. They supply their gills with enough oxygenated water just by swimming with their mouth open.)

The shark's equivalent of arms is its pair of pectoral fins, which are used for steering. They are located on the sides, behind and below the gill slits. The white shark's trademark is its large, triangular first dorsal fin, located on the dorsal (upper) surface. There is an inconspicuous second dorsal fin located farther back, near the base of the tail. A small anal fin is located on the ventral surface, also near the base of the tail. The dorsal and anal fins help stabilize the shark as it moves through the water. The crescent-shaped caudal fin at the rear helps propel the shark through the water. Finally, small, paired pelvic fins are located ventrally, far back on the belly, on either side of the opening for the cloaca, a chamber that collects products of the digestive and reproductive systems before they exit the body. It's easy to tell the sex of any shark: Each of a male's pelvic fins has a long, fingerlike projection called a clasper that is used to introduce sperm into the female's cloaca during breeding. Scientists believe the claspers of a full-grown male megalodon would have been 5 feet (1.5 m) long.

However, scientists later discovered that these model jaws were overly large. The teeth of a great white shark's jaws are not all the same size, as can be seen by closely examining the jaws: On comparison to the teeth in the rear and sides, the front teeth of the jaws are larger. The fossil Megalodon teeth utilized in the \soriginal exhibit model were all huge front teeth. (Remember that more than one shark, not just one, provided the teeth.)

When tooth size variations were taken into account, researchers determined that the Megalodon model's jaws should have only been around 6 feet (1.8 m) wide, which would have meant that the shark was only 45 to 50 feet (13-15 m) long.

Since then, additional fossilized teeth have been found, including a collection of shark teeth. As a result, experts can now anticipate Megalodon's size with greater precision. In reality, a team of researchers under the direction of fossil-hunting paleontologist Michael Gottfried developed a mathematical formula to estimate the size of the original owner of a particular fossil tooth (this formula only applies to the large front teeth). The equation is:

length of megalodon, in meters = (0.96 x [front-tooth height, in centimeters]) – 0.22

The biggest Megalodon tooth discovered so far is nearly 7 inches (18 cm) long. This means the length of the owner of that tooth would have been:

length = (0.96 x 18) – 0.22 = 17.28 – 0.22 = 17.06 meters

This predicted length of roughly 17 meters, or 56 feet, appears to be acceptable; it is roughly in line with the estimation the museum experts made for their diminutive version of Megalodon.

What is the weight of a shark this size? Ichthyologists believe that the weight of a megalodon of the same length would likely be anywhere between 20 tons (18,000 kg) and 40-50 feet (12–15 meters) for whale sharks (Rhincodon typus), the largest of all living sharks.

This was one large fish, weighing close to ten times as much as a truly large white shark and being at least twice as long. That large of a shark must have had a tremendous appetite. What could such a massive animal possibly eat? whale fillet.

A Whale of a Meal

By simply examining a shark's teeth, how can ichthyologists tell what it eats? Because different types of teeth are made to do different things, it's actually pretty simple. Consider the Isurus oxyrinchus, a shortfin mako shark. This swift shark eats fish and squid, both of which are slick and difficult to grasp. The narrow and sharp teeth of a mako are ideal for piercing and grabbing onto such slippery prey.

In order to manage a variety of prey, the hornshark (Heterodontus francisci), another shark, really possesses two quite different types of teeth. The small, pointed teeth on this little shark's front teeth are ideal for grasping and holding onto slick bottom-dwelling fish, while the teeth on its back are stubby and strong, making them ideal for breaking open the shells of sea urchins, crabs, and other crunchy prey. This little shark spends most of its time resting on the seabed.

The strong, pointed, serrated teeth of the great white shark are a horrifying giveaway as to its preferred diet. White sharks, especially adults, prefer to feed on marine mammals, which are typically much too large to be consumed in one mouthful, in contrast to most other sharks, which frequently feed on species small enough to be taken down in one bite. (Elephant seals are a favored diet of the great white; an adult male can weigh as much as 5,000 pounds [2,300 kg]!) Big pieces of meat are frequently pulled from a white shark's massive prey. Its fangs function as a knife and a fork. The teeth's serrated edges cut through flesh and bone like a powerful steak knife, while the teeth's pointed points pierce into a seal's body like a fork's tines puncture a piece of steak.

Because Megalodon's teeth are so similar to the great \swhite's, experts are quite sure it also dined on marine mam╳mals. However, Megalodon likely hunted considerably larger food since most seals would be little more than an appetizer for that "locomotive with a mouth full of butcher blades," whereas the little white shark favors pinnipeds. The most likely main course on the menu would be whales. In actuality, several whale fossil bones display suspicious-looking scratch marks.

The great white shark's teeth were made to catch and consume huge prey. These are the very ominous 20-foot (6-meter) great white's teeth.

ancient Megalodon teeth that have also been discovered may be proof of a bloody conflict between two enormous animals.

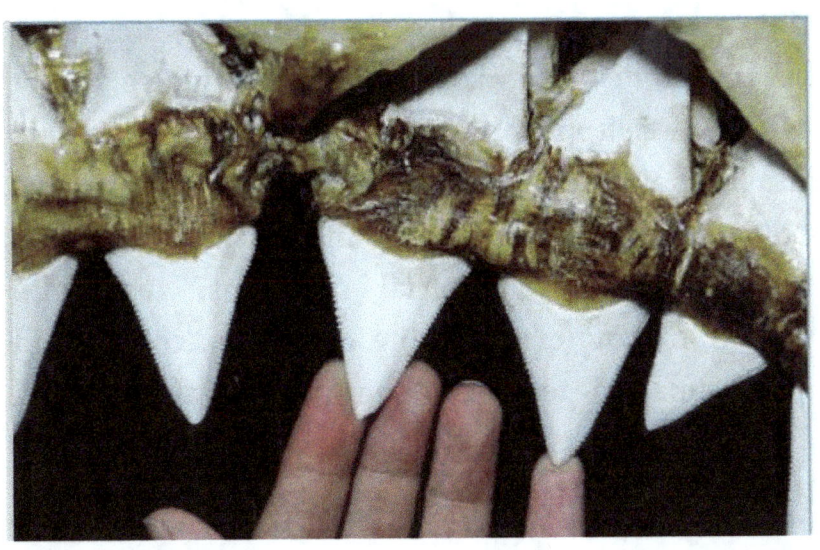

The great white shark's teeth were made to catch and consume huge prey. These are the very ominous 20-foot (6-meter) great white's teeth.

Modern-Day Megalodon Fact or Fiction?

A few persons have claimed to have come with Megalodon-sized monster sharks prowling the seas since 1918. Can we trust these eyewitness accounts? Did these folks witness Megalodon in person, or did they witness another large fish that they mistook for Megalodon? Maybe they didn't see anything at all and simply concocted a number of fantastical fish tales about "the big one that got away."

We can only rely on these eyewitnesses' accounts because none of them offered any conclusive, direct evidence to support their assertions. Therefore, it is vital to look into any information provided by these accounts in order to assess and draw conclusions regarding these potential sightings. It's quite similar to how crime scene detectives collect and assess small pieces of evidence found at the scene of a nefarious act.

It's time to investigate a creature scene!

Story of a Whale of a Fish

There are five eyewitness tales of potential Megalodon sightings that have been documented. All five of the stories will be thoroughly examined in this text. The analysis of two of these accounts, both of which provide startlingly similar descriptions of enormous sharks encountered off the coast of Australia, serves as the investigation's starting point in this part. The first one is a very incredible tale; in fact, it might be the greatest fish tale ever.

The Lobsteredmen's Tale, Case #1

Australian naturalist David G. Stead published this report in his 1963 book Sharks and Rays of Australian Seas. In this instance, Stead tells the tale of a group of Australian lobster fisherman (referred to as "crayfish men" in the narrative) who were tending to their "crayfish pots" when they had an encounter with a large fish. The narrative goes like this:

I wrote down the uproar that the "outside" crayfish fishermen at Port Stephens produced in 1918 when they refused to leave shore for a few days to fish in their usual spots near Broughton Island. The men were working on the deep-water fishing grounds when an enormous shark of almost amazing dimensions appeared, lifting pot after pot filled with numerous crayfishes and stealing, as the men described it, "pots, mooring lines and all." It should be noted that the crayfish pots had a diameter of about 3 feet 6 inches [1 m] and frequently held between two and three dozen good-sized crayfish, each weighing several pounds. The men were in agreement that this shark was unlike anything they had ever imagined. I interrogated a number of the individuals in depth while working with the local fisheries inspector, and they all concurred that the beast was enormous in stature. However, the lengths they provided were generally ludicrous. But the reason I bring them up is to show you the mental state that this peculiar behemoth had left them in. Also keep in mind that these men were accustomed to the water, various weather conditions, and a variety of shark species.

The shark was "at least" three hundred feet long, according to a crew member! Others claimed it was 115 feet [35 m] long, which is how long the wharf was where we were standing. They confirmed that when the fish swam by, a sizable area of water "boiled" over. Although they had all seen passing whales in the water before, this was a large shark. Its terrifying head, which was "at least as long as the roof on the wharf shed at Nelson's Bay," had been seen by them. Obviously impossible! These, however, were fairly stolid and pragmatic men who were not fond of telling "fish stories" or even discussing their catches. Furthermore, they were aware that I, the person they were speaking to, had heard all the fish tales years earlier. One of the things that impressed me was that they all concurred on the big fish's eerie pale appearance.

Stead defines these lobstermen as "prosaic and fairly stolid," indicating that they were cool-headed individuals who were unlikely to be quickly startled or frightened by strange or dangerous conditions. However, whatever these men witnessed while taking care of their lobster traps appears to have disturbed and frightened them to the extent that they delayed returning to their fishing location for a number of days. Stead thinks the lobstermen's claims of the shark's extraordinary length may have been exaggerated accidentally because the normally unflappable men were so enthralled by the spine-tingling excitement of the occasion. But he is certain that these lobstermen witnessed a huge shark unlike any they had ever seen.

This is undoubtedly a whale of a story. The myth has some credence because great white sharks are known to attack buoys and lobster traps. It's even more intriguing because a shockingly similar (and possibly suspiciously similar) tale was recounted several years later.

Case #2: The Tale of the Crawmen

Author B.C. Cartmell wrote the following account in his 1978 book Let's Go Fossil Shark Tooth Hunting:

An 85 foot (26 m) ship encountered engine difficulty in the 1960s around the outer border of Australia's Great Barrier Reef, forcing it to weigh anchor for repairs. The captain and his crew later told friends of seeing an enormous shark that was moving slowly, despite the fact that the sailors eventually refrained from disclosing what they had seen out of fear of public scorn through their ship. They were astounded by its size and its whitish tint. It was at least as long as their boat. Being seasoned sailors, they were equally positive that the creature was not a whale.

This enigmatic creature's incredible size and color make it resemble a somewhat scaled-down version of the monster described in the lobstermen's tale. In the end, it's possible that the lobstermen's depiction of their monstrous shark was correct. However, it is vital to take into account the likelihood that the crew members in Case #2 were simply making up a copycat tale. In any event, assessing both narratives will assist to ascertain whether Megalodon was actually the size and color stated in these two tales.

My, How large your teeth are!

The magnitude of the teeth held by a Megalodon as large as the sharks depicted in Cases #1 and #2 would have astounded Little Red Riding Hood, who was impressed by the size of the Big Bad Wolf's teeth. With a little mathematical trickery, Gottfried's shark-length formula can be changed into a tooth-length formula:

front-tooth height (centimeters) = (length of Mega$\times$lodon [meters] + 0.22) ÷ 0.9

This updated formula can be used to determine the size of the front tooth on any Megalodon specimen. This algorithm yields some rather astounding results when the lengths of the sharks mentioned in Cases #1 and #2 are entered. A shark at least 85 feet long is mentioned in Case #2. Since the formula calculates shark length in meters, the metric equivalent of 85 feet (26 m) must be entered into the equation. If you plug this number into the tooth height formula, you get:

$$\text{front-tooth height} = (26 + 0.22) \div 0.96 = 26.22 \div 0.96$$
$$= 27.31 \text{ centimeters (10.75 inches)}$$

A front shark tooth that was as long as the one in Case #2 would have been around 11 inches (28 cm) long. Wow, that would make a great paper weight. What happened to the shark from Case 1? It was far larger than before. How big would this monster's front teeth be? Estimated length for the smaller shark was 115 feet (35 m). As a result of plugging that into the tooth height calculation, we get:

$$\text{front-tooth height} = (35 + 0.22) \div 0.96 = 35.22 \div 0.96$$
$$= 36.69 \text{ centimeters (14.44 in.)}$$

A tooth 14 inches (36 cm) long would be the diameter
of a pizza! But that's nothing. Just look at the results for the
larger estimate, 300 feet (91 m):

$$\text{front-tooth height} = (91 + 0.22) \div 0.96 = 91.22 \div 0.96$$
$$= 95.02 \text{ centimeters (37.41 in.)}$$

Just think: the front teeth of a 300-foot (91-m) Megalodon would be longer than three feet. To move one of those around, you would need a wheelbarrow.

It is difficult to imagine that any of the shark lengths claimed in the tales of the lobstermen and crewmen are true given the fact that the largest Megalodon tooth ever discovered is less than 7 inches (18 cm) long. It seems conceivable that fossil hunters would have occasionally unearthed teeth much larger than the numerous hand-sized ones that have been recovered so far if Megalodon had ever reached to such enormous sizes. Both of these accounts are tainted by the fact that this is not the case.

A Megalodon tooth is shown next to a great white shark tooth on the left (right). The teeth are comparable in shape and style, although having a wide range in size.

It's as clear as black and White

Let's move on from tooth size. What about the second crucial aspect of the sharks' unearthly white color in these two cases?

This question might be answered by taking a peek at Mighty Tooth's tiny relative. Unbelievably, a shark's feeding habits can be used to accurately forecast its color. A prime example is the white shark. Pinnipeds are the shark's preferred prey, which it chases in the shallows as they leave or return to the security of their rookeries (the beaches and rocky coasts where they congregate when not at sea in search of food). If a white shark wants to have any hope of getting supper, it must be able to sneak up on its prey because pinnipeds are great swimmers and have good eyesight. In order to find prey, the great white uses a cruising feeding technique towards the bottom of the shallows. It stalks up beneath or behind its target when it finds a seal or sea lion in the water, then suddenly leaps forward in a powerful thrust that catches the animal off guard and delivers a massive bite that causes it to bleed to death in a matter of seconds. The shark's dorsal surface is a dark gray, nearly black tint that serves as concealment as the prey fish merges in with the sea floor.

LET'S GET TECHNICAL: GOTTFRIED'S FORMULA

*S*hark expert Michael Gottfried's Megalodon tooth formula is a handy tool. By measuring the height of any fossilized front tooth of a Megalodon, we can determine the body length of its owner. The formula is stated as follows:

length of Megalodon, in meters = (0.96 x [front-tooth height, in centimeters]) – 0.22

For example, to determine the length of a shark whose front tooth is 15 cm long, plug the tooth length into the formula:

length of Megalodon = (0.96 x 15) – 0.22 = 14.4 – 0.22 = 14.2 m (or 46.5 feet)

This formula can also be rearranged to determine how large a front tooth would be on a Megalodon of a specific length. To do this, we simply rewrite the formula.

Start with the original formula:
length = (0.96 x [front-tooth height]) – 0.22
Add 0.22 to both sides of the equation:
length + 0.22 = (0.96 x [front-tooth height]) – 0.22 + 0.22

Since −0.22 + 0.22 = 0, the right side of the equation can be
    simplified, giving:
length + 0.22 = (0.96 x [front-tooth height])
Now, divide both sides of the equation by 0.96:
(length + 0.22) ÷ 0.96 = (0.96 x [front-tooth height]) ÷ 0.96
Since 0.96 ÷ 0.96 = 1, the right side of the equation can be sim-
    plified, giving:
(length + 0.22) ÷ 0.96 = front-tooth height
Finally, flip-flop both sides of the equation:
front-tooth height = (length + 0.22) ÷ 0.96

Suppose you wanted to figure how large a front tooth should
be on that 14.2-meter-long shark. Using this rearranged version of
Gottfried's formula, we get:

front-tooth height, in centimeters = (length of megalodon, in meters
    + 0.22) ÷ 0.96

$$= (14.2 + 0.22) ÷ 0.96 = 15\ cm$$

By using this second form of Gottfried's formula, it's easy to pre-
dict the size of the front teeth of sharks such as the ones described in
the lobstermen's and crewmen's tales.

There is a clear reason why the great white shark appears dark on top and white on the bottom. The shark's colour makes it challenging to see it when it is underwater hunting. When viewed from above, the shark's dark top and the ocean's deep bottom merge into one. The shark's white belly melds with the blue sky when viewed from below.

On the other hand, the ventral surface of the great white is extremely bright, nearly being pure white. The shark's belly provides further concealment when viewed from below, blending in with the brilliant sky. Fish frequently exhibit a color pattern known as countershading, which is characterized by a dark top and a light bottom. Without a physical specimen, scientists can't be certain about Megalodon's hues, but they are reasonably certain that Mighty Tooth, like its smaller, seal-eating sibling, had a dark top and a light belly that allowed it to sneak up on its victim.

Why, then, did the witnesses in Cases #1 and #2 claim that the giant fish was all white in that circumstance? They might have been mistaken: Light reflected from the water's surface could have fooled their sight. Or perhaps this is just more proof that neither of the claims is true. These individuals may have drawn inspiration from a certain classic book authored by Herman Melville in 1851. If these two eyewitness reports are in fact just forgeries, they certainly add weight to the term "white lie," as Moby Dick, the famous sperm whale from Melville, was completely white.

sizing up the occasion

What other reason could the lobstermen in Case #1 have had for making up such a wonderful tale than to pull David Stead's leg? They might have used it as an excuse to hide the unintentional loss of their lobster traps. Because the mooring line to the submerged traps was not tightly secured to its marker buoy, it may have drifted away. The lobstermen wouldn't be able to find their traps the following time they swung by in their boat to harvest their catch if there wasn't that buoy at the surface marking the location of the traps in the deep water below. The lobstermen may have attempted to cover up their error by inventing a tale about an encounter with a monstrous shark that had a sweet desire for shellfish rather than telling the businessman who owned the boat and lobster traps the truth about such a potentially embarrassing and expensive mishap.

Case #2 is equally questionable. Shark expert Ben Roesch claims in a piece for the Cryptozoology Review that Case #2 "drips with tabloid style and reads much like a reworked version of the 1918 gigantic shark." Additionally, he notes that because Cartmell does not provide any sources to support his claims, the account is "useless as evidence" for Megalodon's existence. This would be an instance of hearsay evidence, which has the same value as rumors or gossip.

It's also noteworthy to note that the story wasn't published until 1963, despite Case #1 taking place in 1918.

Case #2 just so happened to take place somewhere "in the 1960s," by coincidence. It's difficult to not be suspicious of the timing of these occurrences, especially since they both allegedly occurred in the Pacific Ocean off the Australian coast (the Great Barrier Reef runs the entire length of the continent's northeast coast, and Port Stephens is on the east coast, a little to the north of Sydney). After the 1918 episode, no one saw any additional enormous, white monster sharks for more than 40 years. The enormous, white monster shark then made a surprise appearance right about the time the story of the lobstermen was published in Stead's book. Australian lobstermen and fishers would have been interested in Stead's extraordinary tale. The timing of these occurrences strongly points to the idea that the crew members in Case #2 simply read or heard Stead's account before deciding to fabricate their own tall tale.

On the other hand, it's possible that a family of enormous white Megalodons lived off the east coast of Australia based on the "fact" that two of these sharks were spotted in the same general area. However, if such were the case, one might anticipate that occasionally a huge tooth would be dragged up from that section of the ocean floor. However, as was already mentioned, no such teeth have ever been discovered.

For many of the shark species that are still alive today, Australia's waters are great hunting grounds. Sharks particularly enjoy hunting off the coast of Australia's northeastern Great Barrier Reef.

First-ever proof: Megalodon

It's pretty obvious that neither Case #1 nor Case #2 makes a compelling case for the contemporary existence of Mighty Tooth in light of the extraordinary sizes and unlikely coloration of the sharks described, the dubious circumstantial evidence surrounding the timing of the publication of Case #1 and the occurrence of Case #2, and the complete lack of any evidence solid enough to sink one's—or Megalodon's—teeth into.

What about the additional eyewitness testimonies?

Perhaps there is further evidence for Megalodon's existence in the tales of a famous novelist and his teenage son. Time to raise anchor, depart Australian territorial seas, and head for the warm open seas of the South Pacific.

Like Father,
Like Son

The only thing that Cases #1 and Case #2 have in common with the following two eyewitness accounts of enormous sharks is that each of them happened in the Pacific Ocean. The sharks mentioned in Cases #3 and #4 resemble one another, but they don't resemble the spectral white giants mentioned in the previous two accounts. Cases #3 and #4 point to an entirely different creature. This warrants a deeper study.

The author's tale is Case #3.

Numerous books by Zane Grey about cowboys and the Wild West are available. In addition, he was a skilled deep-sea angler who enjoyed pursuing large game fish like marlins and swordfish. In Case #3, Grey's encounter with what some believe to have been a live Megalodon is briefly described.

In the 1920s, while deep-sea fishing in the South Pacific, Grey spotted a massive shark swimming close to his boat. The shark, according to his description, was "yellow and green... [with a] square head, immense pectoral fins and a few white spots.... [It was] considerably longer than my boat— conservatively between 35 and 40 feet [10.5 and 12 m].... I figured out that the fish... was not a harmless whale shark but one of the man-eating monsters of Then, I recall being more terrified than I had been in a long time.

Grey implies that he is familiar enough with the whale shark, the largest known living shark, to know that the monster he saw was not one of these calm giants when he describes one enormous shark in this tale. Nevertheless, we must flip every seashell. The whale shark must be ruled out as a potential suspect for Grey's "man-eating monster" in order to reach a reliable conclusion regarding Case #3.

One whale-shaped shark

The whale shark, which is longer than the shark described by Grey, can reach to a length of 50 feet (15 meters), as was previously mentioned. So, the whale shark is still a possibility despite Grey's shark's larger size.

The monstrous shark's head's form doesn't help either. The great white and mako, two predatory sharks that move quickly, have streamlined, pointed snouts, whereas the whale shark has a large, unique blunt snout with a broad mouth at the front that resembles a square with rounded edges. The whale shark feeds slowly on filters. It consumes by gently swimming at the surface while gaping its wide-set jaws open. As a result, water can enter the mouth and exit through the gills, which have filter-like structures that catch small aquatic organisms as they float in the water. Grey claims that the shark he saw had a square-shaped head.

Sharks that swim slowly and eat plankton and algae are called whale sharks. Of all the sharks that are known, it is the biggest. In the open sea close to Donsol, Philippines, a whale shark measuring 20 feet (6 meters) long is feeding.

Grey says his shark had enormous pectoral fins. The relatively big pectorals of whale sharks aid this large-headed shark in steering and balancing itself while swimming. (It's probably no accident that the massive-headed Megalodon that scientists have reconstructed also has comparatively huge pectorals.)

What about the color of the shark? Grey's shark had some white patches and was green and yellow in color. It turns out that the color patterns of whale sharks vary. They have a simple countershaded color pattern with a light belly and a dark background color on the dorsal surface (gray, gray-green, brown, or rusty brown). Usually, there are a lot of white Sharks that swim slowly and eat plankton and algae are called whale sharks. Of all the sharks that are known, it is the biggest. In the open sea close to Donsol, Philippines, a whale shark measuring 20 feet (6 meters) long is feeding. with a checkerboard pattern against the black background color, with yellow spots and vertical stripes on the shark's back. However, the number and shape of spots differ from one animal to another: There are different numbers of spots on different whale sharks. The shark Grey describes has a coloring that is similar to that seen in whale sharks.

This is all quite interesting. Grey appears to have made a mistake. Although he claims otherwise, the enormous shark he describes sounds a lot like a whale shark. It possesses the appropriate body proportions, head form, pectoral fin size, and colour. If Case #4, which also includes the well-known author, leaves any room for dispute, it should prove that the whale shark is the fish that frightened Grey.

Case #4: A Tale of a Teenager

Case #4 happened back in 1933. After a deep-sea fishing trip to the South Pacific island of Tahiti, Grey and his son Loren were traveling back to the United States. They were travelers on board the S.S. Maunganui steamship. Loren, who was a teenager at the time, looked at a splotch of yellowish water visible from the steamer's deck and saw the following scene:

I initially believed it to be a whale, but when the enormous brown tail lifted in the wake of the ship as the fish swam slowly away from the liner, I realized it was actually a huge shark. The enormous spherical head seemed to be at least ten to twelve feet [three to 3.7 meters] across, if not more. This enormous, yellowish, barnacled beast, in my estimation, must have been at least 40 or 50 feet [12 or 15 meters] in length. He was not a whale shark, which has a considerably narrower head and a characteristic appearance of white with huge

42

brown markings. So what was he—possibly a real prehistoric sea monster?

Once more, a large shark described here is comparable in size to either a Megalodon or a whale shark. According to Loren Grey, who claims to know what a whale shark looks like, the large shark he observed was not a whale shark. However, Loren seems to be misinformed about the characteristics of whale sharks, much like his father was: The "white purplish green look," brown dots, and narrow head" that he thinks characterize a whale shark do not represent a whale shark at all.

One seemingly little piece of information, however, actually provides extremely strong proof that Loren Grey's "prehistoric monster of the deep" was a whale shark. Ben Roesch, a shark expert, speculates that the yellow area of water near the shark may have been a "cloud" of yellowish plankton, microscopic organisms drifting in the water. Whale sharks that feed on plankton filters are known to be drawn to such plankton clouds. Additionally, the "yellow and green" hue of his father's shark in Case #3 as well as Loren's shark's peculiar yellow color may be explained by the presence of yellowish plankton in the water. It would appear as though you were viewing the shark through a sheet of yellow cellophane.

Exist any other sharks in the wild that could be confused for the enormous ones the Greys saw? The basking shark (Cetorhinus maximus), which may grow as long as 33 feet, is the only other shark that grows anything near as big as the whale shark (10 m). This is yet another large-headed filter-feeder that moves slowly near the ocean's surface while swimming with its mouth open and sifting water through its enormous gills for small invertebrates. This enormous brown shark has a narrower head than the whale shark, enormous gill slits that run the entire length of the head, and a recognizable cone-shaped snout. Nothing else is likely to be confused with it.

Both the whale shark and the basking shark are filter feeders. After the whale shark, it is the second-largest shark in the world. The basking shark is a calm animal that presents little danger to people.

Second-best practice: MEGALODON

It is very obvious from a careful examination of the accounts of the enormous sharks in Cases #3 and #4 and the circumstances surrounding these encounters that both cases include instances of mistaken identity. Whale sharks, not Megalodon, were most likely the sharks that Zane and Loren Grey saw.

LET'S GET TECHNICAL: FILTER-FEEDING GIANTS

The largest of all sharks—whale sharks and basking sharks—feed on plankton, a mix of microscopic plants and animals, small fish, shrimplike crustaceans and other tiny invertebrates. How can such big animals survive by eating such tiny morsels of food? By eating a load of them: One basking shark caught by scientists had 300 pounds (136 kg) of plankton in its stomach!

Most of the food energy in the ocean is stored in the countless billions of organisms that make up this plankton soup. In order to feed their tremendous bulk, whale sharks and basking sharks must strain thousands of gallons of seawater through built-in filters located in their gills. A feeding whale shark frequently closes its huge mouth and forces seawater back through a soft, spongelike filtering material that collects plankton as the water flows through and out the gills.

This spongy filter is supported by sturdy rods made of cartilage (the same substance that forms the stiff part of your nose) so that it won't collapse and let the plankton escape. The basking shark doesn't even bother to close its mouth when feeding; it just slowly swims forward, mouth wide open, letting the water flow through stiff, bristly, plankton-trapping structures called **gill rakers**. The plankton eventually gets stuck in mucous at the back of the throat, where it is swallowed.

Compared with active, tooth-studded hunters such as the great white and Megalodon, basking sharks and whale sharks lead slow-paced lives of leisure.

The final case is a peculiar one, to put it mildly. The fifth case involves a peculiar shark attack. However, this attack wasn't aimed at anything as appetizing as a dolphin or seal. Not even edible, it wasn't. Unbelievably, a boat was the target of this attack!

Things that Happen at Night

The massive sharks in the first four cases that were looked into did not appear to pay any attention to the boats of the eyewitnesses as they passed by them. The final eyewitness story, which recounts a near contact between a boat and a large fish—possibly a Megalodon—describes a toothy kind of interaction.

The Captain's Tale, Case No. 5

Scientist Ben Roesch has provided the following account of a shark attack on the cutter Rachel Cohen as a brief overview of the intriguing realm of shark behavior. A cutter is a quick patrol boat employed by the coast guard.

Workers discovered 17 teeth imbedded in the ship's wooden hull in March 1954 when it was in a dry dock in Adelaide [an Australian port city]. These teeth apparently resembled those of the white shark. The largest known white shark teeth measure about 6 cm (2.5 in) in height, whereas the reported teeth were reportedly 8 cm (3 in) wide and 10 cm (4 in) high. The "bite" was close to the propeller and the teeth were organized in a semicircle, as is typical of a shark bite, measuring around 2 m (6 ft) in diameter. The shaft of the propeller itself was warped. The skipper of the Rachel Cohen described a tremble the ship felt one night during a storm close to Timor, Indonesia. He believed it had been caused by a collision with a floating tree trunk at the time because they appear to be widespread in the area.

The circumstances of this tale strongly imply that the Rachel Cohen was attacked by a huge shark. Case #5: Could this have actually happened, or is it just another hoax? Fortunately, this narrative gives some incredibly helpful information on the attack and obviously calls for more research.

Propeller shafts and boat hulls: Who's for dinner?

It so happens that shark attacks on boats have been widespread. It was the great white that committed the crime in almost every instance. We can begin to comprehend the minds of the white shark and possibly Megalodon by looking at the specifics of some of these peculiar events.

Shark attacks on boats have been documented throughout history. The great white is typically the offender. In this made-up image from 1908, two voracious sharks are repelled by a group of young girls in a boat.

The apparent insanity of white shark attacks on boats is actually a strategy. Authors Richard Ellis and John E. McCosker offer Ernest Palmer's account of the typical assault strategy used by white sharks that hunt from boats in their book Great White Shark.

The propeller is routinely chewed and rattled by the shark, apparently to test whether the thing is edible. "The first indication of its presence is usually a forceful thump onto the rudder, keel, or side of the boat." Indeed, Palmer saw a shark strike the boat's propeller while he was out fishing, saying, "The propeller struck the shark on the head three times but it continued to pursue till we anchored and captured the shark with three nasty-gashes-in-the-head."

LET'S GET TECHNICAL: SHARK SENSES

Humans rely on five senses to learn about the world around them: sight, hearing, taste, smell, and touch. Sharks have all these senses, but they also have a few others, which they use to monitor their environment in ways people can do only with the help of high-tech instruments.

Sharks have a pair of eyes that are similar to human eyes. Located on the sides of the head, a shark's eyes allow it to see in almost every direction. They are especially good at seeing in the dim light of dawn and dusk, and even at night, when many sharks are on the prowl. This "night vision" is possible because of the presence of a special structure in the back of the eye, the tapitum lucidum, which reflects light onto light-detecting cells in the retina. It's the same structure that makes a cat's eyes glow at night in the beam of a flashlight.

Sharks' ears are located near the top of the skull. Even though the ears have no opening on the surface of the head, they are very good at detecting sounds such as those produced by injured animals flopping and splashing around in the water. Shark ears also have structures called semicircular canals, which help the fish maintain its balance as it moves through the water.

Sharks have a keen sense of smell. The nostrils, located near the tip of the snout, permit water to pass into sensory structures called nasal capsules, where biomolecules such as proteins (for example, oxygen-carrying hemoglobin in the blood) can be detected in concentrations as low as one molecule per one billion water molecules. Sharks can detect the scent of prey miles away.

The most likely explanation for this fairly peculiar behavior is offered by Ellis and McCosker. Sharks have a remarkable variety of senses that they employ to explore their underwater environment.

A shark's taste buds are located on the tongue, allowing it to taste food items as it bites or mouths them. Unusual items found in the stomachs of sharks (for example, bottles and tin cans) are believed to have been accidentally swallowed while being tasted.

Countless nerve cells of various types are located beneath a shark's skin, providing the fish with a very sensitive sense of touch as well the ability to detect bending and stretching of the body as the shark swims through the water. Sharks also possess a lateral line, a network of pressure-sensitive nerve cells called neuromasts, located on the head and along the sides of the shark. By sensing changes in water pressure, the lateral line can detect moving objects. (To get some idea of what this sense is like, climb into a swimming pool and use your hand to swish water toward your leg; even though your hand and leg never touch, your leg can feel the pressure wave of water created by your swishing hand.)

Finally, sharks possess an electrosense that can detect faint electrical fields produced by contracting muscle cells of other animals. These electrical fields are detected by a network of tiny structures, called ampullae of Lorenzini, located in the skin of the head and lower jaw. This sense is used for close-range snooping, as when scrounging around for prey buried under the surface of the sea floor.

The white shark is well-equipped to navigate through its watery world. Megalodon certainly was, too.

Sharks can detect incredibly weak electrical fields, such as those created by the contraction of muscle cells in living things, according to one of their senses, the electrosense. Some sharks have an electrosense that is so acute that it can pick up prey that is buried beneath the sand on the ocean floor. Unexpectedly, a boat's propeller and propeller shaft's metal surfaces also create a weak electrical field in the water. These electrical fields produced by the propeller are thought to be detectable by the white shark, which then bites the propeller to determine whether it is edible.

But that's not the complete picture. Evidence suggests that white sharks don't exclusively attack vessels when they're starving. If they believe their territory, in this case the surf, is being invaded, they may also attack.

Two is a crowd, one is a company.

Off the east coast of Canada in 1953, one of the most well-known white shark attacks took place. On the website of the Shark Research Committee, shark specialist Ralph S. Collier lists numerous shark attacks on boats, including this one:

John D. Burns, a commercial fisherman, went out every day in his dory [a small fishing boat] to catch lobsters with his friend John MacLeod. In search of the valued crustacean, many dories peppered the sea, but only Burns's had a white-painted hull. After leaving the harbor, a sizable shark followed the white-hulled dory for almost a week. The other fishermen continued to observe in shock as the shark prowled behind Burns and MacLeod's dory day after day. As soon as their dory departed the dock, a sizable dorsal fin could be seen behind it. The shark then charged as the dory was sailing by itself on July 9, punching a [8 inch, or] 20-cm hole in the bottom of the boat. After its initial and only strike against the boat, the shark did not come back.

There are two potential theories, while it is obvious that no one knows what the white shark was thinking when it attacked the dory. On the one hand, it's possible that the shark struck the boat in the hopes of consuming it. Then, realizing that the wooden hull was not especially appetizing, it swam away in quest of a more appetizing meal. On the other hand, it's possible that the shark saw the dory as a potential adversary or competitor rather than a prospective food.

The shark ignored all the other fishing boats and focused on Burns's dory. This might be as a result of the fact that it was the only dory with a white hull. Remember that a white shark's lower surface is white. Consequently, it's likely that the shark mistaken the white-bottomed dory for an unwelcome intruder—another white shark. The shark may have been communicating in shark language by ramming the dory, "Get out of here! This is my domain!

Given Collier's account of another shark attack that took place along the California coast in 1989, this interpretation of the shark's actions is not as absurd as it may seem. During this encounter, a great white was seen feasting on a dead harbor seal by a boat with a blue and white hull. The shark left the dead seal and circled the boat, ramming its hull as it did so. Then it repeatedly smacked the boat's propeller and back end with its tail. The boat had by this time drifted away from the seal. The shark went back to its meal and continued to eat.

An key piece of information regarding the shark's motivation for attacking the boat is provided by the tail-slapping activity described in this unusual incident. According to Peter Klimley's research, white sharks will slap their tails to establish a pecking order around a delicious food supply, like the dead harbor seal in the aforementioned story. According to Klimley, "The activity consisted of a shark rising the caudal fin out of the water, stopping to appear as though it were being directed in a certain direction, and then rapidly lowering it while contacting the water with considerable force, sprinkling copious amounts of water in the direction of another white shark on occasion. The shark that makes the biggest splash wins the competition and the prize as two sharks vying for a seal carcass slap water back and forth at each other. White sharks may settle disputes amicably and without fighting or getting hurt by smacking each other's tails.

Collier's account of the great white slapping the boat suggests that it was laying claim to the seal carcass. The shark assumed it had won the "contest," went back to the seal, and started feeding again because the boat made no splashback at all and instead drifted away as though in retreat.

Analysis of the event

How does Case #5 fit into all of this, then? Given the facts shown above about white shark attacks on boats, the captain's story is undoubtedly credible. Boat attacks by great whites do occur. They either bite boat hulls and propellers thinking they are food or thinking they are competitors. (Unfortunately, the information in this tale is insufficient to ascertain the reason why the shark attacked the cutter. It's hardly surprising that no one saw any whale carcasses floating about or tail slapping because the occurrence happened at night during a storm.) What about the other information on the account, though? Do they offer any concrete proof to support the captain's account?

That's a Huge Fish.

The teeth discovered trapped in the cutter's hull, according to the captain's account, belonged to a large shark. Using Gottfried's formula, we can determine how big the shark was, assuming those 4-inch (10-cm) teeth were front teeth:

Megalodon length in meters

= (0.96 x [front-tooth height, centimeters]) – 0.22

= (0.96 x 10) – 0.22

= 9.6 – 0.22

= 9.38 meters, or approximately 31 feet

Fishermen in the Mediterranean Sea managed to land the biggest white shark ever. It measured 23 feet (7 meters). The shark from Case #5 was far bigger than that—it was 8 feet (2.5 m) longer! If it wasn't a megalodon, that shark must have been a truly enormous white shark!

The claimed 6-foot-wide (2 m) bite mark on the hull corresponds with the shark's reported size if those teeth were side teeth because the front teeth would have been significantly longer. (Recall that we earlier mentioned that experts believed a shark with teeth 6

feet wide would be up to 50 feet [15 m] long.)
If only those teeth had been kept!

Third-best practice: MEGALODON

The information provided in Case #5 is totally consistent with the great white shark's known behavior, therefore it cannot be completely discounted as a hoax. However, it's perplexing that no one thought to save any of those allegedly buried 4-inch (10-cm) teeth in the Rachel Cohen's hull. Such a sizable tooth would be a remarkable and priceless discovery.

Actually, a trophy. The absence of teeth as proof raises serious questions about the captain's story. It also does not help that Roesch's sources provide no references that would enable anyone to check and confirm the narrative.

Why would the captain fabricate a tale like that? The Rachel Cohen may have run aground, slamming the propeller shaft on an underwater boulder or coral reef, which was the likely source of the boat's damage. The captain's superiors might discipline him for such an error.

Anyhow, all we actually have is hearsay evidence, and hearsay evidence is useless in creature scene investigations, therefore we must follow Roesch's example and throw out Case #5.

We must take into account the very probable possibility that this majestic animal is extinct because there is no conclusive evidence to support Megalodon's existence. Our investigation's next step will thus be to try to determine what might have caused this shark's death. It will be necessary to travel far back in time, to a time when megalodon still controlled the seas, in order to accomplish this.

Unsolved Mysteries of the Deep Blue Sea

If you have seen the 1993 film Jurassic Park, you may recall the Dr. Grant's jaw-dropping expression of complete awe when he discovers his first living dinosaur. He collapses to the ground, unable to comprehend what he is seeing, and is rendered speechless and weak in the knees. After all, a massive asteroid collision with Earth 65 million years ago was thought to have wiped off the last of the dinosaurs. Everyone is aware that Michael Crichton's Jurassic Park is purely a work of fiction, but does that mean nothing so incredible could ever occur in reality? Is there any possibility that a species believed to have persisted unnoticed for millions of years, leaving no traces—not even a single fossil—and then abruptly reappearing, alive and well, in front of the world's scientists, who were in awe? It turns out that something similar to this did occur a number of years ago: You guessed it! The deep, dark depths of the ocean are home to an animal that was once believed to have vanished even before the dinosaurs were extinct.

What kind of animal performed this incredible disappearance? A fish. It should be noted that this was not a little fish that would be readily missed in the vastness of the ocean. The coelacanth, also known as a 5-foot (1.5 m) long and 127 pounds (58 kg) in weight, was nearly as large as the scientist who first discovered it. It possessed a large, toothy mouth, enormous scales that resembled armor, and short, lobe-shaped fins.

The finding of a living coelacanth was astounding because the fossil record of lobe-finned fishes petered out during the Cretaceous Period, around 80 million years ago.

It's important to look closely at this strange fish tale because many who think Megalodon might still live claim that it is evidenced by the tale of the coelacanth. Perhaps Megalodon could have done the same if this lobe-finned fish could have survived unnoticed and undetected into current times.

Once thought to be extinct, a coelacanth was discovered in the ocean off the coast of South Africa in 1938. Since then, other living specimens have been found, but their sightings are rare, and the fish is classified as an endangered species. Coelacanths are deep-sea creatures that live at depths of up to 2,300 feet (700 m).

The story of Coelacanth

It was 1938, five years after Case #4, that Loren Grey first encountered his "prehistoric monster of the deep." Hendrick Goosen, the captain of the fishing vessel Nerine, delivered a message to Marjorie Courtenay-Latimer, the curator of the East London Museum, after returning to the port city of East London on South Africa's east coast. He extended an invitation for her to visit the dock so she could see the fish that his crew had just brought in with their trawling net. (A trawling net is a huge fish net dragged along the ocean floor.) He informed Courtenay-Latimer that she was welcome to purchase whatever fish she desired for the museum's fish collection.

There was an uncommon fish that Courtenay Latimer had never seen before, in addition to the usual selection of sharks, cod, and other deep-water creatures. It was a big purplish-blue fish with peculiar, lobe-shaped fins and shiny, silvery markings. She purchased the fish and returned it to the museum. She sent a description of the strange fish to James L.B. Smith, a professor at Rhodes University in Grahamstown, some 50 miles (80 km) south of East London, but he was unable to identify it. Smith was a chemistry professor at the university and an authority on fish.

Smith came to the conclusion that Courtenay Latimer was describing a crossopterygian, or lobe-fin, after reviewing the data she had provided. Many researchers hypothesized that this fish was the first vertebrate (animals with backbones) to emerge from the water and move onto land, the ancestor of amphibians. Crossopterygians, like the prehistoric shark Cladoselache, first appeared during the Devonian Period, according to the fossil record. But lobe-fins were believed to be extinct, unlike sharks.

The scientific discovery of the century would be the discovery of a living lobe-fin!

The internal organs were gone, discarded. But when Smith did finally get a good look at the fish's remnants, he instantly realized he was looking at a coelacanth, an actual, genuine crossopterygian! Smith was undoubtedly astounded by what he witnessed, according to Samantha Weinberg, who wrote about it in her book A Fish Caught in Time: The Search for the Coelacanth. Despite my preparation, the initial glimpse left me feeling queasy and strange, and my entire body tingled, he said. I appeared to be turned to stone as I stood. Scale by scale, bone by bone, fin by fin, there was no question in anyone's mind that it was a genuine Coelacanth.

Smith offered some words of wisdom to his fellow scientists in a newspaper story he subsequently wrote about his experience with the coelacanth: "We have in the past thought that we had mastery not only of the land but of the sea. We didn't.

There, life continues much as it did at first. The effect of man is still only a fleeting shadow. This finding suggests that previously thought to be extinct fish-like organisms may still exist in the oceans. Megalodon is thought by some to be one such beast.

The coelacanth is not the only extraordinary, completely unexpected creature to have caught the attention of ichthyologists in the last 70 years.

A fascinating species was unintentionally captured in a U.S. Navy boat's anchor in 1976. Additionally, this was no little guppy-sized fish. This fish was actually a large shark, which is just fitting.

BECOME MEGAMOuTh

When the AFB-14's sea anchors were brought in in 1976, the crew members were in for a rather unpleasant surprise. (A sea anchor is trailed in the water behind a boat; it acts like an underwater parachute, slowing the boat down but not stopping it.) They found that a marine anchor had been devoured by a massive shark with an enormous, obscenely gigantic head as it was drifting 500 feet (152 meters) off the coast of Hawaii. The shark weighed more than 1,500 pounds and measured more than 14 feet (4.3 m) long (680 kg). Its mouth was lined with thick, meaty lips and covered in thousands of tiny teeth.

A coelacanth is a "living fossil." The modern-day coelacanth (its scientific name is *Latimeria chalumnae*) looks almost identical to its fossilized ancestors that lived hundreds of millions of years ago, during the Devonian Period. Its crossopterygian relatives gradually evolved into the first amphibians, some of which eventually evolved into the first reptiles, some of which in turn evolved into the first mammals (the whole process taking a couple hundred million years). The coelacanth itself, however, stayed pretty much just the way it was, a creature obviously very well suited to living the life of a coelacanth.

Coelacanths spend the daylight hours resting in caves hundreds of feet beneath the surface. They are nocturnal hunters, leaving the protection of their caves by night and swimming up into shallower water. There, their large eyes help them find their fish prey in the dimly lit water. Coelacanths eat a wide variety of fishes, including

Because of its enormous mouth, this previously undiscovered fish was appropriately given the name "megamouth," and it later proved out to be another filter-feeding shark. Megachasma pelagios, which means "huge mouth from deep water," is the name given to it by biologists, who later categorized it as a distant relative of the much larger filter-feeding basking shark.

Since the anchor-eater shark was caught in 1976, several additional megamouth sharks have been captured. A 15-footer (4.6 m) was caught in a fisherman's net in 1984 off the coast of California, close to Los Angeles, at a depth of 125 feet (38 m). Others have since been discovered close to Australia and Japan.

Megamouth is thought to be a deep-water filterfeeder, which would account for how long it eluded observation. It would have escaped notice by persons in boats at the surface by spending the majority of its time below the surface.

It wouldn't be tempted to latch onto a giant baitfish on a hook like other open-water predatory fish (like other sharks, marlin, and swordfish) would because it feeds on plankton and other small creatures. This way of existence contrasts sharply with that of Megalodon, the sleek, swift, shallow-water whale-slaughtering machine that was never likely to be kept out of sight for very long.

eels, rays, and small sharks, which they snatch up with their large, powerful mouths. These primitive-looking animals are obviously very capable hunters. Not surprisingly, fishermen of the Comoros Islands usually catch coelacanths on their fishing lines at night, when the coelacanths are out and about searching for prey.

The secretive nature of the coelacanth explains why it remained undiscovered by scientists for so long. If it hadn't been for Marjorie Courtenay-Latimer's lucky discovery of that first coelacanth among all the other ordinary fishes hauled in by the trawler Nerine, we might still be unaware that coelacanths have survived into modern times. This is why some people believe that Megalodon may still be around: It may be living a secretive way of life as well, out of the view of humans—although, quite frankly, it's hard to imagine how 45-foot-long (14 m), whale-eating sharks could go unnoticed for long!

In his book Great White Shark, author Richard Ellis notes that while several megamouths have been spotted so far, no live Megalodons have been seen, as if in rebuttal to those who maintain that Mighty Tooth still exists. One would anticipate seeing Megalodon at least as regularly as the elusive megamouth if it were still alive.

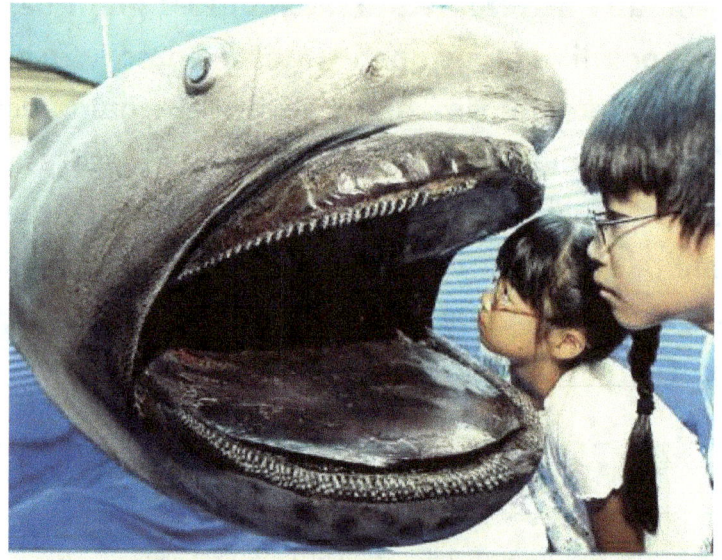

This stuffed megamouth shark was caught in August 2003 in the Pacific Ocean. This one-of-a-kind stuffed fish was on display at a Japanese museum where curious kids could get an up-close look.

DEEP-DIvING?

Recent ichthyological studies have provided some unexpected insights into the migratory patterns of the great white shark, the smaller cousin of the

Megalodon. In shallow coastal areas, particularly near Australia, South Africa, and the east and west coasts of North America, white sharks are thought to spend the majority of their time moving back and forth between feeding and breeding grounds. This belief has persisted for years among ichthyologists. They now understand that this is untrue.

Tipfin, a male white shark previously mentioned, and five other white sharks were given pop-up tags as part of the 2000 study in which ichthyologist Barbara Block of the Tuna Research and Conservation Center, Peter Pyle, and several other colleagues first saw Tipfin.

The researchers found that sharks that visit the Farallon Islands during the elephant seal breeding season stay close to the beach and rarely dive deeper than 90 feet, contrary to what they had previously hypothesized (27 m).

When the seals scattered at the end of the breeding season, something unexpected happened that shocked the scientists: The majority of the sharks with tags moved farther offshore into the Pacific Ocean's deeper waters. Three of the sharks swam to open water several hundred miles to the southwest, while Tipfin, after having a close encounter with the shark-eating orca, swam 2,280 miles (3,650 km) due west, all the way to Hawaii, averaging 43 miles (68 km) per day! Only two of the sharks stayed close to the Farallones. The following November, during the height of the elephant seal breeding season, he then returned to the Farallones.

The researchers found that the white sharks with tags spend up to five months of the year swimming in the deep ocean, frequently at a depth of about 1,000 feet (305 m). Why did those sharks sail across the open ocean for such a great distance? Nobody is certain. One of the academics who worked on the project, Burney Le Boeuf, said, "What they were doing out there is a mystery. Such a lengthy travel may indicate a rendezvous for mating or a change in diet as they were hunting for seals when they were tagged.

Tipfin's journey to and from Hawaii was spectacular, but it was nothing compared to Nicole's record-breaking trek, which she undertook after being tagged in shallow seas close to pinniped rookeries off the coast of South Africa.

The Marine and Coastal Management Department of South Africa scientists tagged 32 white sharks in November 2003, including Nicole. While sharks frequently traveled up and down South Africa's east coast, Nicole headed directly east and swam about 7,000 miles (11,200 km) to Australia! Nicole frequently swam at the astounding depth of more than 3,200 feet throughout her 99-day journey (976 m).

Nicole's pop-up tag broke off and floated to the surface while she was in Australian waters, but researchers know that she later returned to South African seas the following summer: They found her strikingly marked dorsal fin tearing through the waves off the coast of South Africa in August 2004. These findings matched those made the year before.

Some individuals might believe Megalodon is still alive as a result of these white shark migratory research findings. Who is to say that Megalodon doesn't spend a lot of time in the open ocean's deep water, where it has been revealed that Tipfin, Nicole, and other white sharks spend a lot of time hiding from view?

In actuality, nobody can say with absolute certainty that Megalodon does not dwell in the abyss. Maybe we can be 99.99% sure, but we can't be 100% sure until we've looked everywhere in the world's oceans and found nothing. Ellis's logic is nevertheless difficult to contest: if Megalodon were still alive, we would have seen it by now, or at the very least part of its enormous, white teeth.

THE LORD OF THE DEEP

Before we wrap off our discussion of Megalodon, one more issue needs to be addressed. Remember from the introduction that the initial step in every cryptozoological inquiry is to learn as much as you can about the cryptid. Information given by locals who reside in the area where the cryptid hangs out is especially valuable. An illustration of this is the tale of the coelacanth. Despite the fact that until Courtenay-unexpected Latimer's discovery, experts were completely persuaded that coelacanths had long since gone extinct, the lobe-fin was not unknown to the inhabitants of the Comoros Islands, which are located off the east coast of Africa.

Coelacanths are occasionally caught on fishing lines by Comorian fishermen. The lobe-fin is even given a name: gombessa. This instance serves as unequivocal evidence that native people of so-called "primitive" cultures are likely to have a greater understanding of the natural world than highly trained scientists of contemporary, "advanced" cultures from other areas of the world.

Not all animals are found by scientists long after they are known to locals, such as the coelacanth.

Before being discovered by scientists and explorers from Europe and other places, many animals that are now frequently seen running, swimming, or flying in zoos, on the pages of nature magazines, and in TV programs were known to native civilizations for endless years. Examples include the African pygmy hippo in 1849, the gigantic panda in China in 1869, the white-backed tapir from India, a distant relative of the rhinoceros, "found" in 1816, and the mountain gorilla, the largest of all primates, in 1903.

It becomes clear why cryptozoologists start their hunts for cryptids by obtaining as much information as they can from people who live near their target by noting events like these.

What does Megalodon have to do with any of this? Just so happens that the local fisherman on several South Pacific islands think that a shark-like creature that is 100 feet (30 meters) long exists and is known as the Lord of the Deep. It has been a long-held custom for many generations. Even though such a creature would seem impossible to science, one can't help but wonder whether there is more than a tiny bit of truth to this notion. One might also start to wonder if some of the scenarios we looked at weren't quite that unbelievable after all. The accounts given by Zane and Loren Grey seem to involve incidents of identity confusion.

What about the tales told by the crew, the captain, and the lobstermen? The veracity and accuracy of these claims can't be completely ruled out, despite the evidence or lack thereof casting a shadow of skepticism and doubt over them. We weren't there to see what, if anything, truly occurred, after all.

Even though it's a long shot, it's possible that these strange tales about the Lord of the Deep are based on true sightings of that enigmatic creature. The idea is that if the title "Cryptozoologist Captures 'Lord of the Deep'" appears on the top pages of newspapers around the world one day, we shouldn't be too shocked. Such a headline would be accompanied with a picture of a happy individual standing in the sand by the water's edge, close to some beached marine monster, possibly a huge shark, whale, or something completely undiscovered by science. We know from history that such things have taken place before. They probably won't stop happening.

Megalodon, the final report

Now that all the Megalodon data has been gathered and examined, a quick summary of our conclusions is required. The lobstermen's tale (Case #1) tells about an enormous white shark with a liking for lobster that was between 35 and 91 meters (115 to 115 feet) long. Even while marine mammals, particularly whales, are probably this shark's preferred prey, studies of the great white shark, the closest living relative of the megalodon, reveal that Mighty Tooth may in fact make a meal of the lobstermen's catch. Therefore, this portion of the story is credible. But the fish's purported all-white coloring runs counter to what scientists already know about fish color patterns. For cunning predators like the great white shark and, presumably, Megalodon, countershading is more common than not. It would be quite difficult for a monstrously large, all-white shark to sneak up on clever, keen-eyed whales.

The front teeth of a shark the size described in this occurrence were estimated to be up to 3 feet (91 cm) long using Gottfried's calculation. The chance that shark tooth collectors might have missed such enormous teeth would seem exceedingly implausible given that the largest of the hundreds of Megalodon teeth that have been discovered so far is just 7 inches (18 cm) long. The lobstermen's account is a weak source of evidence supporting belief in the modern existence of Megalodon due to issues with the shark's size and color, as well as a potential financial incentive to fabricate such a crazy narrative to explain the expensive loss of the ship owner's lobster traps.

Under careful examination, Case #2's crewmen's story does not fare any better, and for essentially the same reasons. This report is just as hard to believe as Case #1 due to the fish's claimed enormous size and white hue. The timing of the recounting of this story and the release of Case #1 in David Stead's Sharks and Rays of Australian Seas suggests that the crewmen's account is simply a copycat hoax, even if there does not appear to be a financial motivation for making it up.

Cases #3 and #4, the stories of the author and the youngster, are both blatant examples of mistaken identity. Megalodon appeared to be a larger, more muscular counterpart of the great white shark, according to examinations comparing the teeth of the two animals. Whale sharks, which look extremely different from great whites, are described as being remarkably similar to the enormous sharks that Zane Grey and his son Loren saw while out fishing in the South Pacific. The real clincher, though, is Loren's remark of a spot of yellow water close to the shark he observed. The whale shark's preferred meal, plankton, was most likely that yellow smear.

The captain's story (Case #5) is an intriguing narrative since it suggests that a large shark with a 6-foot-wide jaw attacked the cutter Rachel Cohen. The claimed attack on the cutter's propeller is fully consistent with the known behavior of Mighty Tooth's tiny relative, the great white, and that mouth size is acceptable for a shark the size of a huge Megalodon. The incredibly precious, 4-inch-long teeth, which were purportedly discovered trapped in the cutter's hull, do not appear to have been saved by anyone, though. Furthermore, the story's source did not include any references that might be evaluated for veracity and honesty. Consequently, the entire narrative is regrettably just that—a narrative—and nothing else.

In conclusion, we have examined all five eyewitness reports of potential Megalodon sightings. Two of the claims (Cases #3 and #4) are blatant cases of mistaken identity, where the large sharks spotted were almost certainly whale sharks and not "man-eating monsters," while three of the accounts (Cases #1, #2, and #5) are unverifiable at best and hoaxes at worst.

With such scant evidence, it is understandable why the majority of scientists—including many cryptozoologists—believe Megalodon to be extinct. All five trials yielded hearsay evidence, and while some of it corroborated established facts regarding shark size and colour and behavior (such as assaults on lobster traps and boats), other evidence contradicted those facts. Two critical facts cannot be ignored by the information gleaned from these five cases: first, no authentic Megalodon specimen, dead or alive, has ever been discovered; and second, every Megalodon tooth discovered to date is a fossil that is at least 1.5 million years old. The disappearance of the shark's preferred whale prey, fierce competition from the ancestors of the killer whale, and the triple whammy effect of the Ice Age —which led to cooler water temperatures and the disappearance of shallow seas—are all very compelling explanations for Megalodon's likely demise, according to scientists.

egalodon may have disappeared from the world's oceans, but its fossilized teeth can be found all over the world—on the shelves of fossil dealers. Megalodon teeth are among the most popular items purchased by fossil collectors. Prices posted for these

A Christie's auctioneer oversees the auction of a giant *Carcharodon megalodon* shark jaw at Christie's house in Paris on April 7, 2009. The jaw was part of a collection of prehistoric fossils, but the costly fossil did not sell that day.

We must draw the conclusion that there is no reason to think Megalodon still exists—except in wishful imaginations—in light of the aforementioned data. Some people still hold the belief that there is at least one more great big secret hiding in the depths of the deep blue sea: a 50-foot (15 m) locomotive with a mouth full of butcher knives. This is because the tales of the megamouth and the coelacanth demonstrate how adept the ocean is at keeping secrets.

teeth depend on their size, color, and condition, and can range from $10 for a 1.5-inch (3.8-cm) weathered, chipped, and/or cracked specimen, to $7,000 for a 6-inch (15-cm) tooth in pristine condition—having nice, sharp serrations along the edges, and few if any cracks or chips in the enamel.

The enamel in the most valuable specimens has beautiful color patterns that rival those seen in fine jewelry and gemstones. In fact, some dealers polish little Megalodon teeth and mount them as shimmering pendants in unique, eye-catching necklaces.

Fossil dealers usually sell their Megalodon fossils one tooth or necklace at a time, but there are exceptions. In April 2009, the world-famous auction house Christie's conducted its third annual Natural History exhibit and auction in its Paris gallery. Displayed alongside spectacular fossilized skeletons of an ichthyosaur (an extinct marine reptile) and other prehistoric animals were the amazingly lifelike reconstructed jaws of a large Megalodon. (Reconstructed Megalodon jaws are models made of fiberglass.) Adorned with 168 fossil teeth, the huge, wide-open jaws stood 7.2 feet (2.2 m) tall and had an estimated value of well over $200,000.

Although Christie's sold the ichthyosaur skeleton for an impressive $242,652, no one bought the Megalodon jaws; those pricey teeth apparently threatened to take too big a bite out of fossil collectors' wallets.

A Forgotten World

Sharks have existed for a very, very long period. Cladoselache, a little shark that swam the oceans hundreds of millions of years ago, left behind some of the oldest shark fossils. We know this because, close to the southern beaches of Lake Erie, in 350 million to 400 million year old sedimentary rocks known as the Devonian Cleveland shales, paleontologists have found several fossils of Cladoselache. ("Devonian") refers to the period of the geological time scale that spans from 408 million years ago to 360 million years ago. Geologists are scientists who examine rocks.

Although it existed long before the first dinosaurs began to roam the earth, this shark was roughly 3 feet (1 m) long and resembled a normal current shark in appearance. It had the normal array of fins found on modern sharks, particularly the distinctive crescent-shaped caudal fin typical of contemporary speedsters like the mako and the great white. It also had a torpedo-shaped body, a mouth full of pointed teeth, and other characteristics of modern sharks.

Cladoselache was able to swim quickly thanks to its long, aerodynamic body and big tail. It is thought that the extinct ancient shark went extinct about 350 million years ago.

Including Megalodon, Cladoselache is regarded as the ancestor of all sharks. Following Cladoselache's rule, sharks had a period of rapid evolution that resulted in the creation of numerous species with diverse shapes and sizes. However, Cladoselache and the majority of its offspring had one thing in common: they were ferocious predators, able to survive in the same waters as rivals like enormous carnivorous marine reptiles and, later, ferocious whales. (In reality, large fossil fish have been discovered in the stomachs of fossilized Cladoselache specimens, demonstrating that this small shark was in fact an excellent hunter.)

Megalodon is one of the more recent dinosaurs compared to Cladoselache; its fossilized teeth, Carcharodon megalodon, are only about 16 million years old.

Yet Megalodon—unquestionably the top marine predator and lord of the seas—seems to have vanished from the face of the planet. Megalodon fossil teeth that have been found so far are roughly 1.6 million years old. The fossil track then disappears. What possible factors could have led to the extinction of such a strong predator?

THE WORLD IS CHANGING

Off the coast of what would be North Carolina in another 15 million years, it is a warm sunny day. Mesoteras, a mother right whale, prods her calf to the surface so it can take its first breath of air after giving birth. The newborn calf vanishes in a flash of red water when a Carcharodon megalodon's massive dorsal and caudal fins breach the surface of the sea. The Mesoteras calf is killed after being swallowed whole by a 17-meter-long huge [Megalodon] shark, evoking modern-day images of Great White Shark adults consuming seals off the coasts of California and Australia.

John Clay, a paleontologist, detailed this violent incident. Bruner depicts what was likely a typical occurrence in the prehistoric oceans where Megalodon lived. 15 million years.

However, Earth's oceans were very different from what they are today. the current state of things. By examining these variations, we may uncover information that explains the extinction of Great Teeth.

The Earth has been around a long time: 4.5 billion years. To make it easier to study, understand, and communicate about events during this incredibly long history, scientists use the geological time scale (also called the geological time table). This scale divides Earth's history into shorter, more manageable blocks of time. The geological time scale allows geologists to label portions of Earth history in much the same way that taxonomists use their classification system to label organisms. Instead of talking about kingdoms, families, and species, however, geologists talk about eras, periods, and epochs. For example, the shark *Cladoselache* lived during the Devonian Period, which extended from 408 million years ago to 360 million years ago. Megalodon first appeared in the Miocene Epoch, which lasted from 23.7 million years ago to 5.3 million years ago.

The boundaries of the different divisions of the geological time scale represent sudden changes in the types of fossils contained in successive layers of sedimentary rock. (The actual age of these boundaries can be determined by measuring the relative amounts of radioactive elements, such as uranium, contained within the rock layers.) These sudden changes in the fossil record often signify major environmental catastrophes that caused the sudden extinction of many different forms of life. The most famous such mass extinction occurred 65 million years ago, when it is believed that an asteroid smashed into Earth and wiped out the dinosaurs. This catastrophe marks the boundary between the Mesozoic and Cenozoic Eras.

Older rock layers are harder to date accurately than younger ones. Also, the fossil record in older layers is more patchy and incomplete than in younger layers. Nevertheless, the geological time scale is a convenient tool to use when comparing rocks and fossils of different ages.

Geological Time Scale

Eon	Era	Period	Epoch	Age in Millions of Years Before Present
Phanerozoic	Cenozoic	Quaternary	Holocene	Present
				0.01
			Pleistocene	
				1.6
		Neogene	Pliocene	5.3
			Miocene	
				23.7
		Tertiary	Oligocene	
				36.6
		Paleogene	Eocene	
				57.8
			Paleocene	
				66.4
	Mesozoic	Cretaceous		144
		Jurassic		208
		Triassic		245
	Paleozoic	Permian		266
		Pennsylvanian		320
		Mississappian		360
		Devonian		408
		Silurian		438
		Ordovician		505
		Cambrian		570
Precambrian	Proterozoic			2,500
	Archean			3,800
	Hadean			4,550

© Infobase Publishing

The geological time scale divides and subdivides the 4.5 billion years of Earth's history into smaller and smaller units of time.

Sedimentary rocks from areas like North and South Carolina are among the best sources of megalodon teeth in the world. Shark teeth and other items that collected on the surface of the shallow Miocene seafloor helped produce these rocks.

This is not considered to be a coincidence by scientists. They postulate that Megalodon preferred to hang out in these warm, shallow oceans based on the enormous quantity of shark teeth discovered in these rocks. Why? Remember that fossil shark teeth have also been discovered alongside fossil whale bones. Whales from the Miocene era may have likewise enjoyed hanging around in these warm, shallow waters. Modern whale observations offer a theory as to why this was the case.

In warm, shallow waters close to the coast, a variety of whale species give birth. In order to give birth in warm, shallow lagoons along the Californian Pacific coast, female gray whales (Eschrichtius robustus) travel thousands of miles from their rich feeding grounds in freezing arctic seas. They take this action because calves, or newborn whales, lack a significant amount of the fatty, insulating blubber that shields adult whales from the chilly waters of their arctic feeding sites. Calves would swiftly freeze to death if they were born in that chilly water. However, calves born in warm shallows have sufficient time to nurse on their mothers' milk and develop a thick covering of blubber as they gradually move north to arctic feeding grounds. Therefore, it makes sense to assume that hungry Megalodons were there if newborn whales were present in these warm waters.

Another potential explanation for why Megalodons frequented these shallow oceans can be found by examining the predatory behavior of another species of whale hunter, namely humans. Whaling companies used to cruelly exploit gray whales' migratory patterns back in the 1800s. When a female gave birth in a shallow lagoon, the whalers would harpoon the calf from the coast, drag the defenseless baby toward the beach, and wait for the mother to follow. They would then kill the mother quickly when she became caught in the shallow water. Dealing with a mother whale in the open ocean, where she could maneuver easily and maybe ram the whalers' boat or smash it with her tail, was far riskier than doing this.

In Earth's past, there have been a lot of ice ages. Huge glaciers were formed in Canada and the northern United States as a result of the last ice age, which occurred during the Pleistocene Epoch.

Mighty Tooth might have known that mother whales were less maneuverable in shallow water, making hunting calves in shallow water a safer option. After all, it is not necessary to run the risk of getting smashed by a furious mother whale.

Giant sharks likely hunted young whales many times in those shallow Miocene seas, resembling the image Bruner described. Megalodon would have been drawn to whale maternity areas. That's not all, either: Scientists speculate that Megalodon may have given birth in shallow water because female great white sharks, a near relative of Mighty Tooth, do so. (Although many shark species produce eggs, some, like the great white, bear live pups. Newborn white sharks are 1.2 m (4 feet) long miniature replicas of their parents. Baby Megalodons would have been roughly 12 feet [3.5 m] long, according to scientists.) If so, the shallow Miocene waters would have functioned as the enormous shark's nursery and eating area. It makes sense that so many fossil teeth have been discovered in the Southeast of the United States.

Earth's climate changed about 3 million years ago, toward the conclusion of the Pliocene Epoch. All around the planet, temperatures began to fall, and at the poles, particularly the North Pole, where massive glaciers expanded southward, snow and ice started to build in ever-greater proportions. Eventually, vast ice sheets across North America, Greenland, Europe, and Asia covered tens of thousands of square miles of land. The Ice Age was upon us.

The Triple Whammy

The massive northern glaciers, which were towering sheets of ice up to 1.2 miles (2 km) thick, captured so much water during the Pleistocene Epoch's height of the Ice Age that the sea level dropped by more than 330 feet (100 m), completely draining many of the shallow Miocene seas that Megalodon frequented. By this time, the shark had to contend with both the colder water and the loss of its preferred feeding and breeding grounds. That's not all, though. There was now a more recent arrival on the scene, one strong enough to take on the largest, deadliest shark the world had ever seen.

Four killer whales surround their target, a mother and calf gray whale. The calf would finally be killed by the group, but the mother managed to flee.

That's not all, though. There was now a more recent arrival on the scene, one strong enough to take on the largest, deadliest shark the world had ever seen.

This new tough guy hunted in groups, or more specifically, pods, and was strong enough to take down even the biggest whale prey. What kind of person was this new tough guy who was vying with Mighty Tooth for the title of Top Predator of the Seven Seas? None other than the killer whale, Orcinus orca.

It's hard to imagine, but the ancestors of Shamu and Namu, those adorable Sea World clowns who pull divers around the pool, splash audience members in the front row, and kiss small children with their big, fat tongues, may have defeated Megalodon for the top spot in the marine food chain and contributed to the extinction of Mighty Tooth.

Sharks lose a lot of teeth. It's not uncommon for a shark to lose one or more teeth when feeding, especially when the prey has thick or tough skin. A lost tooth, however, is no big deal to a shark. Sharks have several rows of teeth, although only those in the front row are used for biting. The teeth in the other rows are replacements. If a front-row tooth gets yanked out by a struggling seal—or boat hull—the tooth behind it in the second row moves forward to replace it. Sharks have lost so many teeth over the eons that fossilized shark teeth are among the most common of all fossils.

A fossil Megalodon tooth consists of several parts. The relatively soft, spongy center of the tooth is the pulp. (In live sharks, the pulp contains tiny blood vessels and nerve cells.) The pulp is surrounded by a relatively thick layer of a strong, hard substance called dentine. The dentine layer is covered by a thin layer of enamel, a material that is extremely brittle and hard—harder even than bone. The enamel is glued to the underlying dentine by a substance called cementum. The bottom of the tooth, the root, attaches to the jaw. Along the side of the tooth, where the root meets the upper part of the tooth (the serrated, triangular blade), there may be a scarlike mark called the bourrelet. (Megalodon teeth have a bourrelet; white shark teeth do not.)

A tooth lost by a shark settles on the ocean floor and is eventually covered by the muddy, sandy sediment that is deposited in the ocean by rivers. Over time, more and more material accumulates on top of

An adult Megalodon would have been able to compete with a single 25-foot (8 m) killer whale one-on-one. However, it probably didn't have a chance against a pod of numerous intelligent, swift orcas. Orca intelligence may

91

be an understatement. When pursuing prey, these largest dolphins in the family use an incredible range of hunting tactics. It's quite difficult to outwit them.

the tooth. The increasing weight of this ever-deepening layer of mud and sand eventually compresses the material surrounding the tooth into sedimentary rock.

While this compression proceeds, water and chemicals in the sediment slowly interact with the materials in the tooth. As water is slowly absorbed by the tooth, the dentine gradually swells like a sponge, cracking the thin overlying enamel layer, and then slowly dissolves. As the dentine dissolves, it is replaced by various minerals that were dissolved in the water. Meanwhile, the enamel becomes stained by substances in the surrounding sediment, taking on a color ranging from creamy peach or yellow to rusty red, brown, or even black. Only freshly shed shark teeth are white; fossil teeth are always discolored. This is why Richard Ellis and John McCosker comment in their book *Great White Shark*: "Should someone, then, dredge up a *white* Megalodon tooth, we would know that the giant shark became extinct quite recently—or is flourishing somewhere in the vastness of the oceans and has simply lost a tooth."

This entire fossilization process can take up to 100,000 years to complete. Eventually, perhaps millions of years later, sedimentary rock containing a trapped fossil shark tooth may be brought to the earth's surface during mountain-building processes and/or earthquakes, where the rock is weathered by wind and water until the fossil is exposed and, ultimately, detected by the watchful eye of a fossil-hunting cryptozoologist.

Because they hunt in packs and pods, respectively, orcas and wolves share the moniker "wolf of the sea." Orcas surround their prey, herd them into a close circle to prevent escape, and then go in for the kill when hunting schools of fish or pods of their smaller dolphin kin.

The killer whale enjoys eating pinnipeds among other things. In contrast to fish and dolphins, seals and sea lions occasionally have the chance to use an escape maneuver: they can haul out onto land or, in icy climes, floating chunks of sea ice known as ice rafts. (In this case, the word "life raft" may be more suitable.) However, orcas have discovered a number of ways to circumvent this seemingly foolproof escape plan. A hunting killer whale may send surges of water up onto the ice when a seal is trying to hide on an ice raft, washing the seal back into the water where it is rapidly captured and eaten. A seal may be thrown back into the ocean by an orca that occasionally climbs out of the water and onto the ice. Alternatively, it could ram the ice raft from below, knocking it over and sending the seal into the water. An orca may literally hurl itself onto the beach and slide on its belly right up to its startled prey, grab its meal, and then make its way back into the ocean if it spies a seal on land near to the shore.

When a group of orcas hunts a huge prey animal, like a great whale, they surround the target, boxing it in from all sides, and then turn to attack the defenseless victim, chewing off portions of flesh in tandem. When attacking whales that are much larger than orcas, this style of attack is particularly effective. Such aggressive attack on a 60-foot (18.2 m) blue whale has been seen by researchers onboard a Sea World research vessel.

It's plausible that orcas' Pleistocene ancestors utilized this whalehunting tactic against Megalodon, always avoiding the shark's potentially lethal mouth, given sharks are known to be a part of the orca's diet. They may have also defeated the enormous shark by striking it with a battering ram, aiming for the gills or other weak spots.

Even though it was just 12 feet (3.5 m) long at birth, a baby Megalodon would have been a sitting duck for any hungry orca. In reality, humans have seen what might be regarded as a recreation of an orca attack on a young Megalodon. The captain of a tourist boat informed Peter Pyle, a Point Reyes Bird Observatory scientist stationed at the Farallon Islands off the coast of California, that he had just seen a killer whale fight a white shark. When Pyle got on the scene, he observed the smaller of two orcas dragging a dead white shark along the water's surface that was exactly the same size as a newborn Megalodon, measuring 10 to 13 feet (3 to 4 meters). A little while later, the shark's liver squirted out of its battered body, and the orca let go of the body before devouring the liver. The carcass of the killed shark slowly sunk away from view.

Pyle put the puzzle pieces together with his colleague Alisa Schulman-Janiger to identify the likely series of events that resulted in the white shark's demise: One of the orcas had probably just killed a dead sea lion that was floating at the surface (the Farallon Islands are known for their enormous pinniped rookeries). When the hungry shark noticed the sea lion's blood in the water and approached to investigate, one of the orcas became agitated and attacked the shark, killing it most likely by ramming it before shaking it to death.

It was rather unexpected what took place next. Great white sharks frequently prey in the waters near the Farallon Islands from September through December, when elephant seals congregate there to breed. Numerous large sharks typically explore the nearshore seas during these months in search of prey. But the other white sharks rapidly disappeared from the area after the orca killed the curious shark. They must have detected the stench of the deceased shark's torn remains and made the decision to quit their hunting grounds before becoming the orca's dessert. A similar tragedy happened at the Farallones three years later. Researchers looking at white shark migration gave a male white shark by the name of Tipfin a fascinating story. Tipfin's back had an electronic "pop-up" tag affixed by the researchers so they could keep track of the water's temperature and depth while he was swimming. The pop-up tag kept every two-minute temperature and depth record for six months before rising to the surface (thus its name) and sending its data to the researchers via satellite. When the researchers looked at Tipfin's depth data, they were surprised to learn that at nearly the same moment that witnesses had seen a large fish—likely a white shark— that had just been devoured by a killer whale, Tipfin plunged from the surface to a depth of 1,640 feet (500 m) and made a hasty retreat from the islands.

White sharks and orcas obviously do not get along. They are rivals who pursue the same prey, and whales occasionally eat sharks. (White shark attacks on young orcas may also occur on rare instances.) Megalodon and the predecessors of modern killer whales were probably strong rivals for the same reasons. Unfortunately, one species sometimes survives at the expense of the other, which eventually dies extinct, when two species fight for food or other essential resources. Shamu and Namu serve as evidence that the prehistoric orcas survived. The same cannot be said with Mighty Tooth.

The fossil record reveals that many of the whale species that Megalodon likely preyed upon vanished during the Ice Age cool-down, as if orcas weren't a problem for the dinosaur enough. This loss most likely occurred because the whales' natural prey, including fish, squid, plankton, and other organisms, could not endure the colder environment.

It's challenging to pinpoint the exact cause of Megalodon's extinction because of all these events occurring at once. The majority of scientists think that multiple factors combined to dethrone the enormous shark. The absence of Megalodon's favored prey, competition from orcas, and the consequences of the Ice Age are thought by many ichthyologists to have produced a triple whammy that was simply too much for the shark to manage. In any case, the Carcharodon megalodon was vanished by the middle of the Pleistocene. (As was previously mentioned, the oldest Megalodon teeth discovered date back to the Pleistocene, more than one million years ago, providing rather solid evidence that the shark is no longer alive.)

How can palaeontologists know that every Megalodon tooth found is a genuine fossil? That's simple. Shark teeth that have recently shed their enamel are invariably white, whereas fossil teeth have been stained by chemicals in the sediment they were buried in, and can range in hue from a creamy shade to almost pure black. Megalodon teeth have never been discovered to be white.

Despite this, some people continue to hold out hope that Megalodon is still alive and well in the vast waters of the planet. Do these people have valid reasons to believe that Mighty Tooth might still exist, or is this just wishful thinking? The ocean is full of surprises, some of which would be comparable to finding a living Megalodon, even a ghostly white one, as the next section demonstrates.

www.ingramcontent.com/pod-product-compliance
Lightning Source LLC
Chambersburg PA
CBHW070610220526
45467CB00003B/1372